EXPLORING THE SCIENCE OF NATURE

The Nature and Science of
SURVIVAL

Jane Burton and Kim Taylor

Gareth Stevens Publishing
A WORLD ALMANAC EDUCATION GROUP COMPANY

Please visit our web site at: www.garethstevens.com
For a free color catalog describing Gareth Stevens Publishing's list of high-quality books
and multimedia programs, call 1-800-542-2595 (USA) or 1-800-461-9120 (Canada).
Gareth Stevens Publishing's Fax: (414) 332-3567.

Library of Congress Cataloging-in-Publication Data

Burton, Jane.
The nature and science of survival / by Jane Burton and Kim Taylor.
p. cm. — (Exploring the science of nature)
Includes bibliographical references and index.
ISBN 0-8368-2211-0 (lib. bdg.)
1. Natural selection—Juvenile literature. 2. Adaptation (Biology)—Juvenile
literature. [1. Natural selection. 2. Adaptation (Biology).]
I. Taylor, Kim. II. Title.
QH375.B87   2001
576.8'2—dc21   00-063730

First published in North America in 2001 by
Gareth Stevens Publishing
A World Almanac Education Group Company
330 West Olive Street, Suite 100
Milwaukee, Wisconsin 53212 USA

This U.S. edition © 2001 by Gareth Stevens, Inc. Created with original © 2000 by
White Cottage Children's Books. Text © 2000 by Kim Taylor. Photographs © 2000
by Jane Burton, Kim Taylor, and Mark Taylor. The photograph on page 18 (*above*) is by
Robert Burton. The photograph on page 24 (*above*) is by Jan Taylor. Conceived, designed,
and produced by White Cottage Children's Books, 29 Lancaster Park, Richmond,
Surrey TW10 6AB, England. Additional end matter © 2001 by Gareth Stevens, Inc.

The rights of Jane Burton and Kim Taylor to be identified as the authors of this work
have been asserted by them in accordance with the Copyright, Design and Patents
Act 1988. Educational consultant, Jane Weaver; scientific adviser, Dr. Jan Taylor.

Gareth Stevens editors: Barbara J. Behm and Heidi Sjostrom
Cover design: Karen Knutson
Editorial assistant: Diane Laska-Swanke

Printed in the United States of America

1 2 3 4 5 6 7 8 9 05 04 03 02 01

Contents

Words that appear in the glossary are printed in **boldface** type the first time they occur in the text.

Survival of the Fittest

Above: Impalas in Africa rely on grass for food.

The natural world is a complicated place. Animals and plants live together, relying on each other in thousands of different ways. A change in one plant or animal leads to a change in another.

There is not a single living thing on Earth that does not need other living things to survive. Life on Earth is like a huge spider web, where each strand eventually connects to every other strand.

Imagine an antelope grazing for food and munching its way across the African plain. The antelope relies on grass and other plants for nourishment. After it has digested its food, the antelope leaves small dung pellets behind. These, in turn, become food for dung beetles.

Dung beetles are food for crows. The crows need safe places to nest, and they rely on tall trees for this. The trees have flowers and rely on bats to carry the **pollen** to other flowers — and so the **web of life** goes on and on.

Scientists have only just begun to understand the web of life. The vast numbers of plant and animal **species** on Earth and their fascinating lifestyles are all a result of a principle known as survival of the fittest. This principle means that the individuals that are best able to handle current conditions survive and **breed**, and less well-suited individuals do not.

Above: The green dung beetle relies on impala dung to survive.

Opposite: A spider web glistens in the Australian rain forest.

Above: The pied crow eats dung beetles and builds its nest in a tree.

Below: The sausage tree relies on fruit bats to carry its pollen.

Above: A harmless plant bug looks just like the dangerous green tree ant, so **predators** avoid it.

Above: A mantis survives on the forest floor by looking like a dead leaf.

If just a few of the fittest individuals of a plant or animal species can survive, this means that others must die. Death is one of the difficult facts of nature, but it is a natural and necessary process. It is an important part of life.

If living things did not die, there would be no opportunities for new species to develop. Living space for strong, healthy beings would become cramped, and life on Earth would grind to a halt.

Below: This field vole has died, and sexton beetles get ready to bury the carcass.

Below: Orangutans are highly intelligent and survive by using their brains.

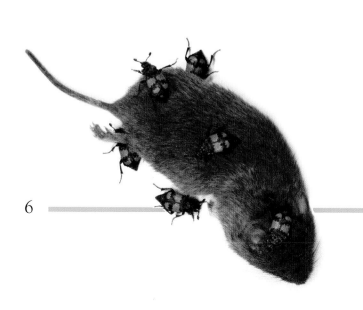

For example, a mother fox might have four cubs, each slightly different from the others. The first one is adventurous. The second one is big and strong, but not very clever. The third one is shy and fearful, while the fourth is small and cunning.

As the four cubs grow up, the first cub wanders far from the den and is eventually run over by a truck. The second cub, instead of running away from a large dog, gets into a fight and is killed. The third cub is too afraid to go anywhere and eventually starves. The fourth cub has the intelligence to search out a good home for itself and, being small, can keep hidden from danger.

The fourth cub is the fittest and survives to breed. Its cubs will tend to be small and cunning, too. Preserving some individuals' best features is a remarkable process called **natural selection.**

Left: A clever red fox survives by moving silently and secretly through the night.

Individual Selection

Top: The European swallow flies between Europe and Africa twice a year in search of food.

In the natural world, if some individuals die, nature can still remain intact. The fittest plants and animals live on to **perpetuate** the species.

In the natural world, what is most important is for the individual plants and animals that *do* survive to produce enough **offspring** to make sure that their species lives on. Many factors help determine success.

Right: Wildebeests regularly trek across the plains of Tanzania to find grass to eat.

Red admiral butterflies are very successful insects. Each spring, thousands of them fly from North Africa and southern Europe to northern Europe, where they breed.

Most of the butterflies raised in northern Europe head south during September when the days in the north become shorter and cooler. During the long and dangerous **migration**, many of the butterflies die. Enough of the strongest fliers get through to produce another generation in the north the following spring.

Each year, just a few red admirals stay behind in the north, where they try to **hibernate**. In the past, nearly all of them died from the cold. Recent warmer winters in northern Europe have allowed a few red admirals to survive hibernation. This is a way that species can develop through the selection of individuals.

Left: On its way north in spring, a red admiral butterfly rests on the seashore. It flew across the English Channel.

Above: Before heading south in autumn, a red admiral butterfly feeds on **nectar** from ivy flowers.

Below: While it is winter in Europe, the European bee-eater lives among elephants in Africa.

Species Selection

Top: The common frog is an excellent survivor. It is usually pale brown in color.

Above: To survive cold winters, a Hermann's tortoise must hibernate.

Nature is much more interested in an entire species than in individuals. If an individual dies, it will be replaced by other individuals of the same species. If the *species itself* dies, however, it is lost forever and it becomes **extinct**. The history of life on Earth includes a history of extinctions. It is possible that as many as ninety-nine percent of all the species of animals and plants that have ever lived are now extinct.

Most of these extinctions have occurred naturally, but humans are becoming increasingly

Right: This bear is actually an **albino** black bear. Albinos do not survive well in the wild.

Above: Golden tadpoles sometimes appear in nature, but they are rare.

Above: Golden tadpoles grow into golden frogs, but they do not survive well.

responsible. For instance, at one time, passenger pigeons where so numerous in North America that their flocks darkened the skies for days at a time. Then Europeans arrived with their guns and shot the pigeons by the millions. The last passenger pigeon died in a zoo in 1914.

Natural extinctions sometimes result from changes in Earth's climate. Species that cannot **adapt** to these changes become extinct. To survive increasingly colder weather, for instance, **mammals** must grow thicker coats and **reptiles** must learn to hibernate. Over several thousand years, a species may change so much that it becomes an entirely new species.

For a species to remain unchanged for millions of years is somewhat rare. Normally, species **evolve** gradually into other species, which are better suited to the current conditions. The natural extinction of a species may take millions of years.

Success Stories

Top: A painted lady butterfly rests on a thistle flower.

Above: The seed heads of dandelions can be found in most countries of the world.

Opposite: Barn owls hunt mice in Europe, Asia, North America, Africa, and many other parts of the world.

Some species of plants and animals are so adaptable that they are able to live almost anywhere in the world. When a species is found throughout the world, it is said to be cosmopolitan.

Barn owls are very cosmopolitan birds. They can be found throughout Europe and Asia. They live in Africa and in North America. Barn owls can survive in so many parts of the world because their requirements for survival are so simple. The owls need only **uncultivated** grasslands to hunt for mice and voles and hollow trees or buildings to build nests.

The painted lady butterfly is a cosmopolitan insect with simple requirements. Painted ladies need warm sunshine to fly in, flower-nectar to feed on, and thistles or nettles for their caterpillars to eat. Almost anywhere in the world where these conditions exist, painted ladies or their very close relatives can be found.

In the plant world, dandelions have spread to most regions. Dandelions need only bare ground for their seeds to grow and a reasonable amount of rain to keep the soil moist.

Truly successful species will always survive somewhere in the world, even when conditions change at a rapid pace.

Quick-Change Artists

Top: Rock doves once nested only on cliffs. The ones that nested on city buildings, which are similar to cliffs, became known as pigeons.

Some species of plants and animals survive in a changing world because they are able to change fairly quickly.

During the **Industrial Revolution** in Europe, there was so much smoke in the air that the trunks of trees in many areas turned black. Pale-colored moths, called the peppered moth, rested on tree trunks.

These pale-colored moths were easily spotted on the blackened trees by birds that ate them. Soon, an all-black race of peppered moths appeared in the world. Birds could not see the black moths on the sooty tree trunks, and so the species survived.

Right: When the trunks of trees in and around cities became blackened with smoke during the Industrial Revolution, peppered moths were easily spotted by birds. The occasional black moth survived better, and soon black moths were more common than pale ones.

Left: Black rats have changed their habits from living in jungles to living in cities. They can now be found throughout the world.

Thousands of species of rats live in forests, fields, mountains, and rivers. Only a few rat species have changed their lifestyles in order to live in buildings with people. Two rat species have reached all parts of the world because they are able to travel as stowaways on trains, trucks, and ships.

Black, or ship, rats have spread to most cities in warmer parts of the world. These rats eat almost anything from insects and grain to bones, candles, or soap. They live along rocky shores, in trees, and in buildings. Brown or common rats are perhaps even greater survivors than black rats. They are at home in places that are wet, even in sewers.

Below: Cliff-nesting herring gulls have discovered that buildings also make good nesting sites.

Living in a Crowd

Top: Fiddler crabs live in burrows on mud flats near the sea. Males fight each other to keep their burrows.

Survival is not just a case of having enough food and shelter. For many living things, survival means being able to **compete** with others of the same species. Animals will fight each other to get what they need. Even plants battle for living space.

Flamingos are large birds that nest in colonies. At nesting time, they crowd together, with each pair of flamingos defending a small patch of ground around their nest. The survival of a flamingo chick is entirely dependent on its parents' ability to keep other flamingos from trampling the nest.

Many species of fish crowd together in shoals, or groups, for their own protection. One fish on its

Right: Greater flamingo nests are close together, and parents squabble to protect their chicks.

Left: Trees in a tropical rain forest compete for light. Those that grow fastest or tallest survive.

Below: These characins gather in a school for protection. When it comes to feeding, however, they compete with each other.

own is easy for a predator to kill, but a school of fish darting in all directions is confusing to the predator. When it comes to feeding, however, all the fish in the school must compete for the same food. The individuals that survive are the ones that are best at competing in a crowd.

In the plant world, individuals of some species compete with each other and also with other species. They do this by spreading their leaves over the ground so they can get most of the light. The plants that survive are the ones that grow the fastest because of the extra light. Plant roots also do battle underground over minerals and water.

When Conditions Get Tough

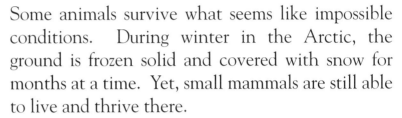

Top: Emperor penguins survive in the toughest conditions on Earth. The males keep the eggs warm during the long, dark Antarctic winter.

Above: A layer of down and a layer of fat protect king penguin chicks from the cold, Antarctic temperatures.

Right: An organ pipe cactus survives desert conditions by storing water in its thick stems.

Some animals survive what seems like impossible conditions. During winter in the Arctic, the ground is frozen solid and covered with snow for months at a time. Yet, small mammals are still able to live and thrive there.

Lemmings and dwarf hamsters burrow beneath the snow, where they are sheltered from the wind and the worst of the cold. Underneath the snow, these little animals are invisible to predators, such as arctic foxes. The lemmings and hamsters survive by eating frozen leaves and plant shoots.

Hot, dry deserts provide animals with tough conditions of the other extreme. Small animals that live in deserts need to be able to escape the midday heat and, above all, exist with very little water. Even such large animals as elephants can survive in deserts if they know where to find water. Over time, elephants in southwestern Africa have learned where to find water holes. During the long, dry season, they travel great distances between water and feeding areas. If young elephants did not learn this routine from their parents and remember it, they would not survive.

Dry conditions occur in other locations as well. **Droughts** sometimes affect places that normally have plenty of rain.

When the soil starts to dry out during a drought, earthworms burrow deep into the ground. They dig small round chambers in which they tightly coil their still-moist bodies. The coiled bodies preserve moisture, and the worms wait for rain to come again.

Above: The desert elephants of southwestern Africa survive by knowing where to find water during the lengthy dry season.

Below: Earthworms survive a drought by burrowing deep into the ground and conserving the moisture in their skins.

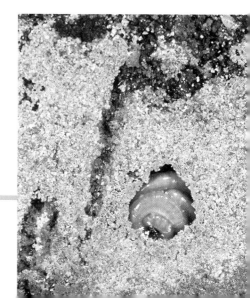

Above: Brine shrimp live in salty water. If the water dries up, the eggs survive for years, blowing around in the air.

Natural catastrophes can happen anywhere in the world. Forests are destroyed by fire. Lakes and ponds dry up. Floods sweep over the land, drowning everything in their path. Many animals and plants die in these catastrophes, but enough survive so that the species can continue.

Brine shrimp live in salty puddles. When a puddle dries up in the baking sun, all the shrimp in it die. It is a catastrophe for the inhabitants of the puddle, and there may be years of drought before rain refills the spot. When it does, however, new shrimp appear as if by magic.

The secret of the brine shrimp's survival lies in their remarkable eggs. The eggs are carried in sacs by the females. When all the adults die in the drying puddles, the eggs are scattered in every direction. Brine shrimp eggs are as small as grains of dust and can blow around in the desert winds for

Right: Brine shrimp even survive the heat and dryness of Death Valley in California and Nevada.

Above: This ohia lehua bush has survived in a crack in the rocks of a volcano.

Above: A fern struggles through the hard rock of a dried **lava** flow.

as long as twenty years. Even after many years, a rainstorm creates the perfect conditions for the eggs to hatch. Brine shrimp eggs can survive blistering heat and total **desiccation**. Very few living things are able to do this.

Another place where conditions are tough is a rocky seashore that is pounded by waves. Even there, certain animals survive. Limpets cling to the rocks with their strong suckerlike feet. Between the waves, crabs scuttle from one refuge to another.

Left: Goose barnacles attached to seashore rocks can survive violent pounding by waves.

21

Those That Don't Make It

Top: Deinonychus probably became extinct about 100 million years ago. It ate animals that fed on plants.

Above: Triceratops was a plant-eater that may have become extinct because Earth's climate became colder.

Right: The large Triceratops died out 65 million years ago, but small, furry mammals survived.

When conditions get *too* tough, certain animal and plant species may die out. Climate changes can create conditions to which some species just cannot adapt. Predators may come along and be so successful at eating a species that eventually no individuals are left.

Perhaps most importantly, similar species may evolve that compete for food and living space with existing species. This competition drives the original species into extinction.

Some scientists believe that the extinction of the dinosaurs sixty-five million years ago happened as a result of the Earth's climate suddenly becoming

colder. The plants on which the **herbivorous** dinosaurs fed may not have been able to grow in the cold. The dinosaurs had no fur to insulate them against the low temperatures. Fur and feathers may have allowed the earliest mammals and birds to survive longer than the dinosaurs.

The main cause of recent extinctions has been human activity. The introduction of animals from Europe and other areas to Australia and New Zealand has been disastrous for native species there. When foxes, cats, rats, and weasels were brought from Europe, the **marsupials** and birds in Australia and New Zealand proved easy **prey** for them. Introduced rabbits, goats, donkeys, and sheep ate the plants that once only native herbivores ate.

Above: The weka is a rare, ground-living bird. It lives only in New Zealand and is in danger of becoming extinct because it is easy prey for cats, rats, weasels, and other mammals that were introduced from Europe.

23

Survival Through the Ages

Top: The very first mammals on Earth were long-bodied, short-legged predators like this common genet.

Above: This peripatus looks like a worm with legs. It has survived, almost unchanged, for many millions of years.

Right: The tuatara has survived for millions of years. It now lives isolated on a few small islands off the coast of New Zealand.

Some creatures and some plants have survived almost unchanged for many millions of years. In the damp forests of Australia, New Zealand, South Africa, and Central America, a strange creature moves over the ground on many unjointed legs. It is not a worm, and it is not an insect. It is a peripatus. Peripatus began life on Earth over 400 million years ago and has changed very little over time. Peripatus has survived such a long time because similar damp forests have always been available to it.

Off the coast of New Zealand, there are islands on which a large lizard-like animal lives. It is not closely related to the lizards found in the rest of the world. This animal is a tuatara. Two hundred million years ago, reptiles similar to the tuatara

were common. Now, only a few of them survive. They owe their survival to their **isolation**. There are no predators and little competition from other species on the islands where they live. Tuataras are now a protected species.

The ginkgo tree has grown throughout the world for over two hundred million years. Chinese priests may have saved the tree from extinction by planting it around their sacred shrines.

Living in a Human World

Top: There are very few areas left in North America where herds of bison can roam.

Above: To get its fur, poachers have hunted the sea otter almost to extinction. The sea otter is now protected.

All of the richest and most of the poorest land on Earth has been **cultivated** to grow food for ever-increasing numbers of people. Cultivation means destruction and death to wild plants and animals.

Countless thousands of species have become extinct during the last hundred years. The rate of extinctions is increasing all the time as forests are cut down and grasslands are plowed or grazed by domesticated cattle.

Some plants and animals survive by adapting to the human world. For instance, some seed-eating birds and small mammals have changed their diets from wild plant seeds to grain crops. However, the larger animals, such as tigers, wolves, rhinoceroses, and elephants, are losing their habitats. Without their homes, they cannot survive in the wild.

Right: Wolves need a lot of space and large prey. There are just a few places left in the wild where wolves are able to live and hunt.

Left: The black rhinoceros is illegally hunted for its horn. Very few of these magnificent animals are left in the wild, but their numbers are increasing due to improved protections.

Only recently have many people begun to realize that humans are part of the web of life, dependent on the natural world for survival. The more plant and animal species that become extinct, the less likely humans are to survive.

More and more steps are being taken to preserve the natural world. For instance, scientists are collecting and storing the seeds of rare plants. Citizens are voicing their concerns over protecting natural habitats. This is a battle that must be won.

Left: Poachers kill African elephants for their ivory tusks. Authorities are trying to control the illegal ivory trade. In some areas, the elephants are beginning to make a comeback.

 # Activities:

The Natural World

What things do *you* need to survive? What are the essentials that you cannot live without? In the modern Western world, most people do not often have to think about surviving on a daily basis. Stores are stocked with food. In fact, many of us eat too much. We can easily buy warm clothes. Drinking water comes out of faucets in our homes, which are heated to comfortable temperatures in winter. Because we do not have to think about survival, we often forget just how frail humans are. Plan an expedition into uninhabited country. You will discover that survival is no simple matter. It requires a great deal of thought and effort.

Expedition in the Wild

Going on an expedition into wild country is an exciting and daring thing to do. You will be hungry and thirsty at times. You will often be uncomfortable and tired, but the rewards are great. You will learn what the natural world is like. It is a far different

place than the unnatural world of cities and towns that people have created.

To be part of an expedition, first discuss it with your parents or guardian. Then contact a local scouting organization to find out if there is a camping expedition coming up that you can join.

There are three basic requirements for human survival — food, water, and shelter. How many of these items can you carry with you on the expedition? How many of them can you find when you reach your campsite?

Is there water at your destination, and is it safe to drink? Are there fish in the water that you can catch and eat? Are there materials you can use to build a shelter?

You should try to get the answers to all of these questions before you start off on your expedition.

In addition to thinking about your own needs, you must also know something about the requirements of the animals you may encounter. Bears can be dangerous, especially when they have cubs. Mosquitoes will want to bite you. Be prepared for anything and everything that can happen.

Survival in the City

Knowing the habits of animals in the natural world is just the opposite of knowing the habits of animals in the city.

It is quite surprising to discover just how many animals have made homes for themselves among the bricks and concrete of towns and cities.

Birds do fairly well in cities. In developed areas, there may seem to be only sparrows and pigeons, but there can sometimes be gulls and falcons. In parks and gardens, there may be many different species of birds.

To observe and learn about city birds, go to the library and get a bird book for your region. Also bring along a notebook, pencil, and a pair of binoculars (*see page 28*).

Keep watch for birds wherever you go in the city. Write down the ones you see and where you see them. Also make a note of what the birds are doing when you see them. Are they feeding, and, if so, on what? Are people feeding them, or are they finding food on their own?

If you see a bird carrying food or nesting material, watch it carefully. With luck, you may see its nest.

Mini-Survivors

Small, ground-dwelling creatures, such as spiders, woodlice, centipedes, and insects survive quite successfully in cities. To study them, you will need a notebook, pencil, and a good pair of eyes. Parks and gardens are the best place to look for these creatures. Inform the park guides that you are on a nature study. Gently turn over stones and logs, and make a note of what you find underneath. Be sure to carefully replace any item that you disturb.

Many tiny city creatures are active only at night. One way to view some of them is to set up a container in the ground. Use a straight-sided, wide-necked container (*see above*). Choose an out-of-the-way site, where the earth is soft. Dig a small hole. Place the container into the hole so that its top is level with, or slightly below, the surface of the ground. Visit the site each morning, and identify any visitors. Always be sure to carefully release the creatures back into the wild.

Below: A young orangutan learns survival tactics from its mother.

Glossary

adapt: to undergo changes to better suit changed circumstances.

albino: a condition where an animal or human has no color in its body.

breed (v): to mate for the purpose of producing young.

compete: to strive against others in order to get something or to survive.

cultivated: tilled and worked soil so that crops can be produced from the soil.

desiccation: drying out to the point where no water remains.

droughts: long periods of time without any rain.

evolve: to change and develop gradually.

extinct: no longer existing.

herbivorous: eating only plant material.

hibernate: to spend winter in a state of rest.

Industrial Revolution: the period of time in history when a transition from manual labor to machine labor took place. The Industrial Revolution began in Britain in 1760 and in the United States in the last part of the nineteenth century.

isolation: solitude; the state of being completely alone.

lava: molten rock.

mammals: warm-blooded, furry animals that produce milk to nourish the young.

marsupials: mammals that carry their young in a pouch on their bodies.

migration: a journey from one region to another.

natural selection: the process in which animals and plants best suited to their environment survive to reproduce.

nectar: the sweet liquid produced by flowers. Bees, birds, and other animals feed on this liquid.

offspring: young plants or animals produced by their parents.

perpetuate: to take part so that a species can last indefinitely.

pollen: male cells produced by flowers in the form of fine grains. The grains are usually yellow in color.

predators: animals that hunt other animals for food.

prey: animals that are hunted by other animals for food.

reptiles: vertebrates that breathe air, including alligators, crocodiles, lizards, snakes, and turtles. These animals crawl or move on their bellies or on their small, short legs.

species: a biologically distinct kind of animal or plant. Similar species are grouped into the same genus.

uncultivated: not tilled and worked (such as the soil) but left in a natural state.

web of life: the relationships between all living things on Earth.

Plants and Animals

The common names of plants and animals vary from language to language. Their scientific names, based on Greek or Latin words, are the same the world over. Each kind of plant or animal has two scientific names. The first name is called the genus. It starts with a capital letter. The second name is the species name. It starts with a small letter.

African elephant (*Loxodonta africana*) — Africa 19, 27

albino black bear (*Ursus americanus*) — North America 10

barn owl (*Tyto alba*) — worldwide 12

black (ship) rat (*Rattus rattus*) — worldwide 15

brine shrimp (*Artemia salina*) — worldwide 20

dandelion (*Taraxacum officinale*) — worldwide 12

earthworm (*Lumbricus species*) — Europe 19

greater flamingo (*Phoenicopterus ruber*) — Africa, India, southern Europe 16

green tree ant (*Oecophylla smaragdina*) — Australia 6

Hermann's tortoise (*Testudo hermanni*) — southern Europe 10

impala (*Aepyceros melampus*) — Africa 5

king penguin (*Aptenodytes patagonicus*) — Antarctica 18

ohia lehua bush (*Metrosideros collina*) — Hawaii 21

orangutan (*Pongo pygmaeus*) — Borneo, Sumatra cover, 6, 29

organ pipe cactus (*Stenocereus thurberi*) — North America 18

peppered moth (*Biston betularia*) — Europe, Asia 14

pied crow (*Corvus albus*) — Africa 5

rain forest, tropical — Australia 17

red admiral butterfly (*Vanessa atalanta*) — Europe, Asia 9

red fox (*Vulpes vulpes*) — North America, Europe 7

sexton beetle (*Necrophorus vespilloides*) — Europe 6

weka (*Gallirallus australis*) — New Zealand 23

Books to Read

Animals in the Fall (Preparing for Winter).
  Gail Saunders-Smith (Pebble Books)
How Animals Protect Themselves.
  Animal Survival (series). Michel Barré
  (Gareth Stevens)
In Peril (series). Barbara J. Behm and
  Jean-Christophe Balouet
  (Gareth Stevens)

Rain Forest: Lush Tropical Paradise.
  Wonderworks of Nature (series).
  Jenny Wood (Gareth Stevens)
The Science of Animals. Living Science
  (series). Lauri Seidlitz (Gareth Stevens)
The Thermal Warriors: Strategies of Insect
  Survival. Bernd Heinrich (Harvard
  University Press)

Videos and Web Sites

Videos

Animal Adaptation. The Wonderful World of Animals. (Simitar Video)

Animal Colors. Amazing Animals Video. (DK Vision)

Birds and Migration. (International Film Bureau)

Secret Weapons and Great Escapes. (National Geographic)

Web Sites

www.redcube.nl/bos/

www.janegoodall.org/rs/rs_history.html

www.tusk.org/

www.nationalgeographic.com/dinorama/frame.html

www.worldwildlife.org/windows/pennies/index.html

www.earthlife.net/begin.html

Some web sites stay current longer than others. For further web sites, use your search engines to locate the following topics: *adaptation, animal survival, migration, predators,* and *prey.*

Index